M103 Heavy Tank

A VISUAL HISTORY OF THE AMERICA'S ONLY OPERATIONAL HEAVY TANK 1950-1970

Published by
Ampersand Group, Inc.
A HobbyLink Japan company
235 NE 6th Ave., Suite B
Delray Beach, FL 33483-5543
561-266-9686 • 561-26609786 Fax
www.ampersandpubco.com • www.hlj.com

Acknowledgements:
This book would not have been possible without the considerable help of Tom Kailbourn, Chris Hughes, Scott Taylor, Don Moriarty, Pat Stansell, Mike Kalbfleisch, the late Jacques Littlefield, the staff of the Patton Museum, Scott Hamric and the 3 ACR Museum, the National Archives, and the TACOM LCMC History Office and of course my wonderful wife Denise.

Sources:
Report of the Army Ground Forces Review Board, 20 June 1945

Report of the War Department Equipment Board, 19 January 1946.

Notes on Development Type Material, 120mm Gun Tank T43, 1 October 1951.

"Firepower," R. P. Hunnicutt, Presido Press, Novato, Ca, 1988.

Front cover: Lessons learned during tests of the pilot T43s led to many changes to the production T43s during their assembly at the Chrysler plant in Newark, Delaware. Because of these many revisions, the production vehicles were redesignated the 120mm Gun Tank T43E1. After an initial order for 80 examples, a total of 300 of these tanks were produced, ultimately redesignated the M103 Heavy Tank. Seen here is T43E1 U.S. Army registration number 30170205, with U.S. Army Ordnance markings on the side of the turret. (NARA)

Title page: An M103A1 of C Company, 1st Tank Battalion, 1st Marine Division, comes ashore off a floating causeway at Camp Pendleton during Amphibious Landing Exercise (PHIBLEX) Swan Dive on 15 February 1960. The code C-21 is marked on the bow and the turret. (NARA)

Rear cover: Tests of the T43s and M103s indicated that these vehicles suffered from subpar sighting systems. To remedy this shortcoming, the Army modified two T43E1s, installing turret baskets and a new T52 stereoscopic rangefinder and T33 ballistic computer to be operated by the commander, not the gunner, and moving the gunner's station forward in the right side of the turret, providing him with a direct-sighting telescope and a repositioned gunner's periscope. Although the Army did not convert any more T43E2s, the Marines converted 219 of their T43E1s to T43E2s, such as this one, and these were standardized as the 120mm Gun Tank M103A1. (NARA)

©2015, Ampersand Publishing Group, Inc. All rights reserved. This publication may not be reproduced in part or in whole without written permission from the publisher, except in cases where quotations are needed for reviews.

Table of Contents

T43 5

T43E1, M103 17

T43E2, M103A1 55

M103A2 65

M51 Tank Retriever 106

The Heavy Tank T29 was designed in early 1945 to counter the increasingly heavy tanks Germany was producing, and work continued on it after the war to meet the threat of the Soviets' formidable tank arsenal. The T29 featured a 105mm main gun. (TACOM LCMC History Office)

Introduction

During WWII, the Allied forces encountered German tanks of considerable size and power, with impressive firepower and armor protection. Each Allied nation engaged in development of weapons to counter these Axis monsters. But, immediately following the end of WWII, the former Allies found themselves to have ideological, and potentially wartime foes. Tank designs once initiated to combat a common foe were now refined to counter each other.

In the US, the T43 heavy tank was built to counter to the Russian IS-3 heavy tanks. Fortunately, the two adversaries never faced off. Ordnance Committee Minutes (OCM) 32530 from 1 December 1948 outlined the desired characteristics of the new 120mm gun-armed tank, which had an estimated weight of 58 tons. Power was to be supplied by an 810 horsepower Continental air-cooled engine, and a four-man crew was planned. As the design was refined, numerous changes were made, resulting in revised military characteristics for the vehicle to be issued under OCM 33333 on 24 April 1950. Among the changes instituted at that time was the inclusion of a fifth crew member, the increase of turret ring size from 80 to 85-inches, and most noticeably, the adoption of elliptically shaped hull and turret castings. The first pilot T43 was built at the Detroit Arsenal, with assembly beginning in October 1950, and the vehicle being completed in June of the following year. The second pilot was also produced by the Arsenal, but delivered to Chrysler, who installed redesigned turret components to overcome deficiencies discovered during initial testing. Two more pilots were built in Detroit, with the final two pilots, numbers 5 and 6, being produced by Chrysler in Newark, Delaware.

The new heavy tank's hull strongly resembled the concurrently developed M48 medium tank, albeit considerably oversized. The heavy tank had one more road wheel than the M48 and had six return rollers per side. Unfortunately, beyond styling, it also shared the powerplant of its smaller sibling, resulting in a heavy tank that was automotively underpowered. Not only was the air-cooled V-12 underpowered for the massive tank, limiting top speed to 25 miles per hour, but it was also incredibly thirsty. The tank's 268 gallon fuel capacity giving it a range of only 80 miles. The heavy tank turret, however, was all-new and huge in order to house its 120mm main gun and its tremendous recoil. Unlike most tanks, the powerful 120mm gun used two-part ammunition, with separate powder and projectile. For this reason there were two loaders in the turret, with the commander and gunner seated in the very large turret bustle.

The powerful new tank, which had been created as the "Heavy Tank T43," was redesignated on 7 November 1950, becoming "120mm gun tank T43" following an army-wide policy of designating tanks not by weight, but by armament.

Delivered in 1947, the Heavy Tank T30 pilots had the same chassis as the Heavy Tank T29 but had a 155mm main gun instead of the 105mm gun and a Continental AV1790 V-12 air-cooled engine, whereas the T29 had a Ford GAA V-8 engine. (TACOM LCMC History Office)

Tracks removed, a Heavy Tank T30E1 is prepared for rail shipment in October 1949. The T30E1 was used as a testbed for an automatic loading system for the 155mm gun, since the ammunition had proved very difficult for a crewman to handle.

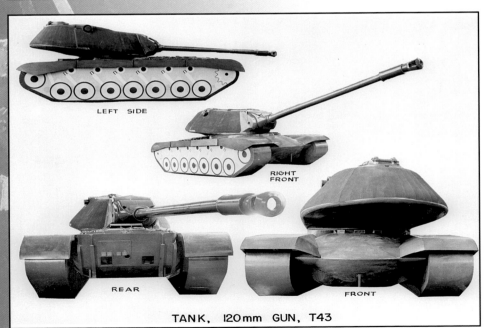

TANK, 120mm GUN, T43

Top left: The Heavy Tank T32, photographed in March 1946, used many components of the M26 Pershing tank to produce an extended-length, enhanced-armor vehicle powered by a Ford GAC 12-cylinder engine. The main gun was a 90mm T15E2. **Top right:** The two experimental Heavy Tank T34s were based on the Heavy Tank T30 chassis and armed with the T53 120mm gun, originally designed as an antiaircraft weapon. A large armored plate was on the rear of the turret bustle to counterbalance the long gun barrel. **Above left:** During the Korean War, the U.S. Army embarked on a crash program to field a heavy tank to be designated the T43. Shown here is a Detroit Arsenal model of the vehicle dated 19 January 1951, five months before the first pilot was completed. **Above right:** This 12 September 1950 Detroit Arsenal model depicted an early M103 concept featuring elliptical hull and turret castings. It also included seven pairs of road wheels on each side, along with six pairs of return rollers. (TACOM LCMC History Office, all)

The T43

The prototype of the M103 series of heavy tanks was the Heavy Tank T43, which was in effect a scaled-down version of the U.S. T34 Heavy Tank, the 1945 experimental design combining a 120mm gun and the T30 chassis. With its smoothly curved turret and elliptical bow, from certain angles the T43 bore more than a passing resemblance to the later M48 Patton tank. Shown here is the first pilot T43, built at the Detroit Arsenal from October 1950 to June 1951, following which it was delivered to Aberdeen Proving Ground, Maryland, for testing. It had an unloaded weight of 110,000 pounds and carried a crew of five. (Patton Museum)

The first pilot T43 has a dust cover over the muzzle in a photo dated 21 June 1951. The three-rung ladder was welded to the side of the turret. Above and to the rear of the ladder is the dome-shaped cover for the right objective of the rangefinder. (Patton Museum)

Top left: The turret of the first pilot T43 is traversed to the rear. At the top is the rear of the commander's cupola. Prominent casting marks are at the lower center of the bow. The U.S. Army registration number, 30163674, is in white near the top of the bow. **Top right:** On the first pilot T43 as seen from the left rear, at the top of the rear plate of the hull are lifting eyes and the collapsible travel lock for the 120mm gun in its stowed position. The right final-drive housing is in view next to the track on the right side. **Right:** The flat roof of the turret of the first pilot T43 is seen from above with the gun pointing to the front at Aberdeen Proving Ground on 7 July 1951. To the front of the commander's cupola is the loader's hatch cover, fitted with counterbalance springs. (Patton Museum, all)

Top left: The 120mm gun barrel secured to the travel lock is observed from another angle at Aberdeen Proving Ground in July 1951. The small, raised panels on the rear deck were exhaust deflectors, and were to each side of the muffler cover. **Top right:** The 120mm gun of the first pilot T43 is traversed toward the right rear quarter, and the bipod-type travel lock is the production T43/M103 heavy tanks. **Above:** In a photograph taken during a round of engineering and endurance tests at Aberdeen Proving Ground on 7 July 1951, the turret of the first pilot T43 is traversed to the rear, and the barrel of the 120mm gun is secured in the travel lock. (Patton Museum, all)

Top left: On the rear of the hull on the inboard side of the left tail light assembly was the armored cover for the external telephone control box. Below the box are two square transmission access plates. Farther below are two round plates for accessing the brake rods. **Top right:** In a view of the rear of the third pilot T43 as secured for transport on a flatcar, details of the used on the pilot T43 heavy tanks were 28 inches wide, with a pitch (front-to-rear length) of 6.94 inches. **Above:** The second pilot Heavy Tank T43, 30163675, was built by the Detroit Arsenal, with Chrysler installing revised turret components. Unlike the first pilot M43, the second pilot had a bore evacuator on the barrel of the 120mm gun aft of the muzzle brake.

This is part of a series of 25 March 1952 Aberdeen Proving Ground photos demonstrating the method of blocking the third pilot T43, 301630676, for rail shipment. Steel rods were used as stays for securing the vehicle to the flatcar; the rods fastened to the front of the turret had turnbuckles. (Patton Museum)

The third pilot T43 is viewed from the front with the turret to the rear on the flatcar. In the right background is an M26A1 tractor. The Detroit Arsenal designed and built the first three pilot T43s. Subsequently, Chrysler assumed design duties and manufacturing for T43 pilots four through six. (Patton Museum)

Details of the travel lock and the rear of the hull of the third pilot Heavy Tank T43 are visible in this view of the tank secured to a flatcar. Prominent on the rear of the hull is the external telephone box. Between the turnbuckles over the rear deck are the exhaust deflectors, with X-shaped stiffeners. (Patton Museum)

One of the original T43 prototypes photographed at Chrysler, Newark in 1954. This tank lacks its mantlet cover but the two tubes for the coaxial .30-caliber machine guns are evident. (NARA)

The fifth pilot of the Heavy Tank T43 was delivered to the U.S. Marine Corps for evaluation, shown here next to a Medium Tank M48. Although some sources assert the Marines received the sixth pilot T43, TACOM records indicate it was the fifth pilot. (NARA)

The USMC pilot T43's turret is traversed to the rear and the main gun is secured to the bipod travel lock. The machine gun mount on the commander's cupola has been removed in this photo and the following one. A T-shaped muzzle brake is present on the main gun. (NARA)

The same USMC pilot T43 is viewed from the rear at Quantico, Virginia, with the travel lock still in the raised position. Although the Army was not satisfied with the tests of the T43s, the Marines were very much interested in acquiring a force of these tanks. (NARA)

The T43E1

Lessons learned during tests of the pilot T43s led to many changes to the production T43s during their assembly at the Chrysler plant in Newark, Delaware. Because of these many revisions, the production vehicles were redesignated the 120mm Gun Tank T43E1. After an initial order for 80 examples, a total of 300 of these tanks, which were eventually redesignated the M103 Heavy Tank, were produced. Seen here is T43E1 U.S. Army registration number 30170205, with U.S. Army Ordnance markings on the side of the turret. (NARA)

On 4 January 1950 the US Army ordered 80 production vehicles, with improvements dictated by testing of the pilots. The vehicles would be produced by Chrysler in Newark, and were later designated T43E1. The army's new heavy tank was of considerable interest to the Marine Corps, who had felt their tanks were obsolete, and heavy support would be critical in future operations. The outbreak of the war in Korea mid-year solidified this thinking, and on 20 December 1950 the Navy ordered for the Marines 195 of the new tanks, valued at a half-million dollars each. This order was later upped to 220 vehicles.

Regrettably, problems with the new design primarily related to main gun fire control continued to persist, and rather than being delivered, the production vehicles were placed in storage at the plant pending resolution. Testing revealed 114 deficiencies needing correction in order to make the vehicles suitable for troop testing. Ultimately, the army opted to only implement 98 of these, with the improvements being applied to only 74 of their 80 vehicles, and the modified tanks were standardized as M103. The army M103s were dispatched initially to Fort Hood. The six unmodified vehicles were retained for further testing.

General Data

Model	M103	M103A1	M103A2
Weight*	125,750 lbs	125,750 lbs	128,000 lbs
Length**	445.5"	445.5"	445.5"
Width	148"	148"	148"
Height**	113.375"	113.375"	113.375"
Track	115"	115"	115"
Armor, Hull Front	4.5" lower, 5" upper	4.5" lower, 5" upper	4.5" lower, 5" upper
Armor, Gun Shield	4" to 10"	4" to 10"	4" to 10"
Armor, Turret Sides	2.75" to 5.375"	2.75" to 5.375"	2.75" to 5.375"
Crew	5	5	5
Max Speed	25 MPH	25 MPH	23 MPH
Fuel Capacity	268 gallons	268 gallons	440 gallons
Range	80 miles	80 miles	300 miles
Electrical	24 Volt negative	24 Volt negative	24 Volt negative
Engine	AV-1790 series***	AV-1790 series***	AVDS-1790-2A
Displacement (In3)	1791.7	1791.7	1791.7
Engine Type	90-degree V-12, air cooled	90-degree V-12, air cooled	90-degree V-12, air cooled
Fuel	Gasoline	Gasoline	Diesel
Horsepower	810 @ 2800 RPM	810 @ 2800 RPM	750 @ 2400 RPM
Torque	1600 ft-lb @ 2200 RPM	1600 ft-lb @ 2200 RPM	1710 ft-lb @ 1800 RPM
Transmission	CD-850-4A or CD-850-4B	CD-850-4A or CD-850-4B	CD-850-6 or CD-850-6A
Crossdrive Speeds	2 Forward, 1 Reverse	2 Forward, 1 Reverse	2 Forward, 1 Reverse
Turning Radius	Pivot	Pivot	Pivot
Track Type	T96, T97, T97E1, T97E2	T96, T97, T97E1, T97E2	T96, T97, T97E1, T97E2
Armament Main	120mm M58	120mm M58	120mm M58
Secondary, Coaxial	2x .30-caliber M1919A4	1x .30-caliber M37	1x .30-caliber M37
Secondary, Flexible	1x .50-caliber M2	1x .50-caliber M2	1x .50-caliber M2
Ammunition Main	33 rounds	33 rounds	33 rounds
Ammunition .50-Cal.	900 rounds	1,000 rounds	1,000 rounds
Ammunition .30-Cal.	8,150 rounds	5,250 rounds	5,250 rounds

A Heavy Tank T43E1 with U.S. Army Ordnance markings on the turret is loaded on a 60-ton tank-transporter semi-trailer hitched to an XM194E1 8x8 15-ton tractor in a photograph dated 9 September 1953. The trailer had 12 wheels in three tiers. (TACOM LCMC History Office)

The U.S. Army planned to deploy this combination of vehicle types as its tank force in the mid-1950s, ranging from heavy to light tanks. From left to right they are the T43 heavy tank, the M48 and M47 medium tanks, and the M41 light tank. (Rock Island Arsenal Museum)

*Measured with main gun facing forward and anti-aircraft machine gun mounted. **Fighting weight.
***Engine models used include AV-1790-5B, AV-1790-7, AV-1790-7B, AV-1790-7C. Displacement (In3)

A T43E1 undergoes tests at Fort Knox, Kentucky, on 24 August 1955. The main armament was the 120mm Gun T123E1 (standardized as the 120mm Gun M58). Tests of the T43E1 in 1955 found it to be unacceptable, and most of the fleet went into storage. (NARA)

Top: T43E1 U.S.A. number 30170196 displays its 120mm gun at maximum elevation of 15 degrees. Elevation was achieved by either an electro-hydraulic mechanism or by manual operation. This also applied to the traverse operation of the turret. **Bottom:** The same T43E1 is shown with the main gun at minimum elevation of -8 degrees. The 120mm gun of the T43E1 had a maximum rate of elevation of four degrees per second: thus, it could complete its entire range of elevation in about six seconds. (Patton Museum, both)

T43E1 registration number 30170196 is observed from the front right at Fort Knox. The .50-caliber machine gun on the mount on the front of the commander's cupola could be controlled remotely by the commander within the turret. Below the gun is a vane sight. (Patton Museum)

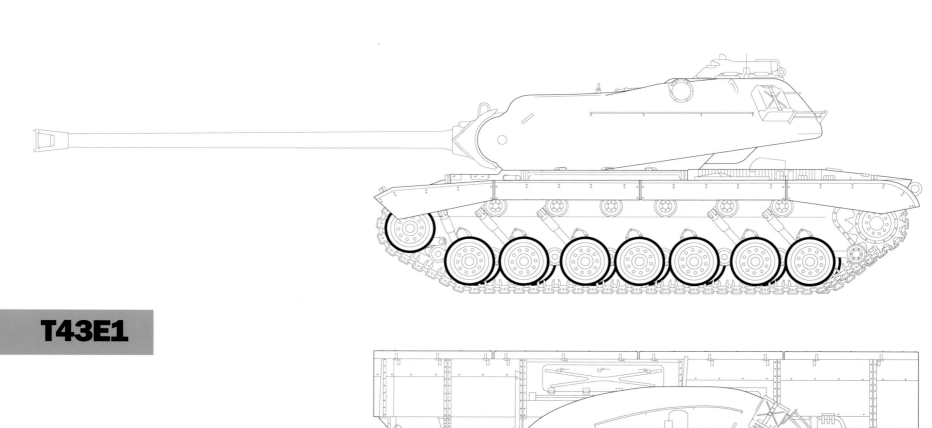

T43E1

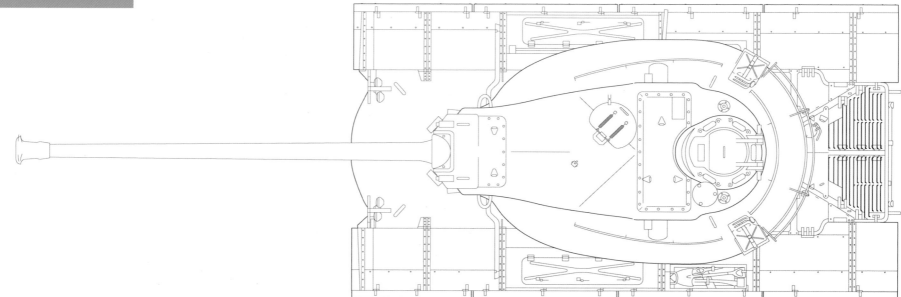

T43E1 30170196 is seen from the rear, with the external telephone box and the travel lock in its stowed position prominent on the rear of the hull. On the rear of the turret is a bulge for the turret ventilator, with the cap on top. "AFF BD 2" and "130" are on the fenders. (Patton Museum)

Top left: The same T43E1 seen in the preceding photo is viewed from the front. The T43E1s had a different design of headlight brush guards compared to the pilot T43s, with a curved upper piece. Marked on the bow is "AFP. BD. NO. 2 / TEST OPERATION." **Top right:** Like the T43 pilots, the T43E1 had a torsion-bar suspension with seven individually sprung dual road wheels on each side. The wheels mounted size 26-6 rubber tires. Shock absorbers were provided for the first three and the last two road wheels on each side. **Above left:** In an aerial view of a T43E1, to the front right of the commander's cupola is the housing for the gunner's sight. The gunner was positioned well to the rear on the right side of the turret. To the front of the sight housing are the cover for the rangefinder and the hatch. **Above right:** On a test course a Heavy Tank T43E1 drives up a steep slope. This model of tank could negotiate a maximum grade of 60 percent. In addition, it could drive over a trench up to 7.5 feet wide and could surmount a vertical wall with a height not to exceed 27 inches. (Patton Museum, all)

T43E1 registration number 30170196 climbs a 60 percent grade during trials for Board No. 2, Continental Army Command, at Fort Knox, Kentucky. A large protractor with a gravity-operated pointer is attached to the fender to indicate the angle of the slope. (Patton Museum)

The same T43E1 undertakes a shallow-fording operation during tests at Fort Knox. The T43E1 had a maximum specified fording depth of 48 inches. The water is not up to the tops of the road wheels yet. The 120mm gun of this tank had a T-shaped muzzle brake. (Patton Museum)

During the shallow-fording test, the T43E1 crosses a stream at Fort Knox. The water is approximately even with the tops of the road wheels. A seldom-used deepwater fording kit was available that permitted fording in water up to the depth of the turret roof. (Patton Museum)

Following a shallow fording of a creek at Fort Knox, a T43E1 ascends a moss-covered dam, displaying its ability to surmount a slippery slope. At the rear of the hull, the right side of the travel lock is slightly raised, showing the curved yoke and latch at the top. (Patton Museum)

The hull is awash as the T43E1 operates in 58 inches of mud: ten inches deeper than the specified 48-inch shallow-fording depth of the vehicle. At a depth of 48 inches, the water level would be approximately at the vertical centers of the track-return rollers. (Patton Museum)

At Fort Knox, a T43E1 comes ashore after a successful shallow fording of 45 inches, just three inches shy of its maximum shallow fording depth. The shore here is extremely muddy, and the chevron tracks were churning it up in an effort to gain purchase. (Patton Museum)

A T43E1 tank, U.S.A. number 9A1950, is parked with the turret traversed to the rear and the 120mm gun secured to the travel lock at Rock Island Arsenal in early May 1955. Adjacent to the T43E1 is a T48 or M48 medium tank. In the left background is the clock tower of the arsenal. (Rock Island Arsenal Museum)

The 120mm main gun of T43E1 U.S. Army registration number 9A1950 has been elevated from the travel lock during a demonstration at Rock Island Arsenal on 5 May 1955. Elements on the turret include a ventilator, bulges for antenna mounts, a liquid-container holder, the right rangefinder-objective hood, a stowage basket, and hand rails. (TACOM LCMC History Office)

A T43E1 climbs a paved 40 percent slope at Aberdeen Proving Ground on 10 September 1953. This area featured various grades for testing vehicles' climbing limits. To the left is a sign with arrows indicating a 60 percent grade (foreground) and a 50 percent grade. "U.S. ARMY / ORDNANCE" is marked in white on the turret side. The muzzle brake is the early cylindrical one with a round hole on each side. The tank commander is standing in his cupola hatch and is wearing a yellow tanker's helmet. (NARA)

Left: A different T43E1 than the one in the preceding photo takes its turn climbing the 40 percent slope at Aberdeen Proving Ground on 10 September 1953. This slope translates to 21.8 degrees. The maximum grade the T43E1 was capable of was a 60 percent slope, which

tration number 30170205. **Right:** A T43E1 at Aberdeen Proving Ground in May 1954 has its turret traversed to the rear, displaying a revised travel lock with a tripod instead of a bipod design. The two rear legs were still attached to swivel mounts on the upper rear of the hull, and a

Above: A commander stands atop the cupola of a T43E1 at Aberdeen Proving Ground on 11 May 1954. The bulges on the rear of the turret bustle are for, from left to right, the turret ventilator and two radio antenna mounts. Only the right antenna is installed. To the front of the right antenna mount is a five-gallon liquid container. **Top right:** A closer view is of- brush guards of the headlights mirrored the semi-elliptical shape of the turret. On the front edge of the bow are two lifting eyes with tow shackles. **Above right:** A T43E1, apparently USA number 30170205, wades trough a water tank at Aberdeen Proving Ground on 10 September 1953. The vehicle appears to be proceeding at its specified shallow-fording

During vehicular tests, a T43E1 climbs a vertical wall. This tank could climb a maximum vertical wall of 36 inches. This wall appears to have been lower than the maximum. This photo offers a good idea of the manner in which the road wheels were able to flex. (Patton Museum)

A T18 bulldozer assembly has been installed on the bow of T43E1 30170201 in a photograph dated 18 January 1955. Such an installation could enable an attack force to clear obstructions while enjoying the protection offered by a heavy tank. (Patton Museum)

Top: The same T43E1 with a T18 bulldozer assembly shown in the preceding photo is viewed from the left side, with some of the hydraulic mechanisms, hoses, and ribbing on the rear of the blade visible. The travel lock for the 120mm gun is the bipod type. (Patton Museum)

Above: The T18 bulldozer-equipped T43E1 is observed from the right side in another 18 January 1955 photograph. The headlight arrays were raised to clear the top of the blade, and the brush guards were redesigned, with flattened instead of curved tops. (Patton Museum)

A rear view of T43E1 30170201 offers a clear view of features on the vehicle, such as the external telephone box, the tail light assemblies with their hooded guards, the travel lock, treads, and the muffler housing. To each side of the main gun was a coaxial machine gun. (Patton Museum)

After deficiencies in the T43E1s disclosed during testing had been fixed with 98 separate modifications, the tanks were standardized as the 120mm Gun Tank M103 in early 1956. The main gun was standardized as the 120mm Gun M58 in the Mount M89. (Patton Museum)

An M103 appears in a 9 June 1958 photo. The M103 retained the remote-controlled .50-caliber machine gun mount on the front of the commander's cupola. A total of 74 of the Army's T43E1s were upgraded to M103 standards; the other six remained test vehicles. (Patton Museum)

Gun muzzles almost touching, two M103s of D Company, 2nd Battalion, 33rd Armor Regiment, are at the gunnery range at the Bergen-Hohne training area in Germany in May 1959. "DEBRA" is painted on the left turret and "DESIRE" on the right turret. (NARA)

A "62" bridge-classification sign is on the bow of this M103 of D Company, 2nd Battalion, 33rd Armor Regiment. The M103 retained a .30-caliber coaxial machine gun on each side of the main gun. The gunner had to do his sighting through a periscope. (NARA)

A jettison fuel kit fabricated from welded tubes and two 55-gallon drums was mounted on the rear of this M103. This kit extended the M103's usual range on paved roads from 80 miles to 145 miles. In the background, the engine-compartment grilles are open. (Patton Museum)

As mentioned, the M103 was initially consigned to the 899th Tank Battalion when sent to Europe. The 899th was later designated 2nd Battalion, 33rd Armor, but in the late fifties and early sixties, companies of heavy tanks were often subordinated to other units. In this case, an M103A1 of the 33rd Armor is serving with the 34th Armored Regiment during winter maneuvers in 1960. The turret is traversed to the rear in this shot and perhaps even in the cold weather it was too warm for the commander when he was located over the rear exhaust. (NARA)

A jettison fuel kit with the drum not installed is visible from the side on this M103 of D Company, 2nd Battalion, 33rd Armor, at Grafenwöhr, Germany, on 15 September 1959. The name on the side of the turret is indistinct but appears to be "DEROGATORY." (NARA)

This shot was probably taken the following winter during operation Wintershield II. At this time the 33rd Armor was part of the 7th Army. Again the tank is traveling with the turret traversed. This photograph provides an excellent view of the heat shield installed underneath the turret. Barracks bags line the turret stowage basket. (NARA)

A column of M103s of D Company, 2nd Battalion, 33rd Armor, pauses along the entrance to Hahnbach, Germany. The closest tank has the name "DEMON" painted on the turret. The forward leg of the travel lock is not engaged on this tank, but is on the second tank. (Don Moriarty)

A fascinating view of the M103 at the end of its service life in the U.S. Army in the summer of 1962, this shot also depicts the only variant of the M103, the M51 tank retriever. Note the boom in the middle distance. Many of the vehicles seen here are being prepared for long-term storage. The M47s in the background have already been partially preserved. The M103A1s in the foreground had been most recently in service with the 34th Armored Regiment. The 34th had been attached to the 24th Infantry Division from May 1960 to June 1962. (NARA)

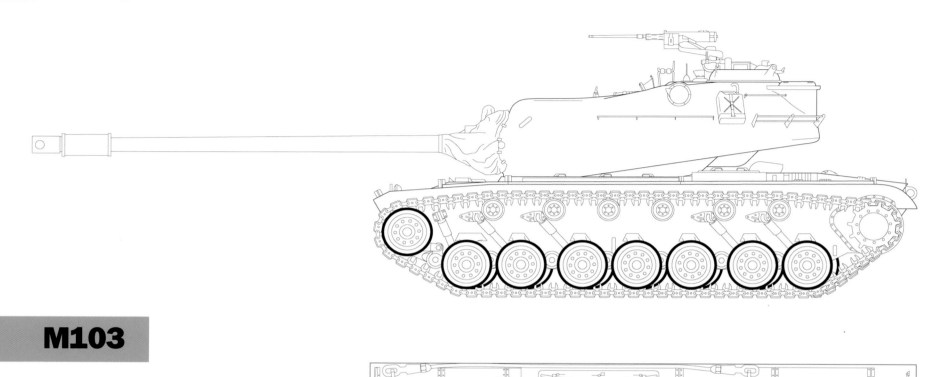

M103

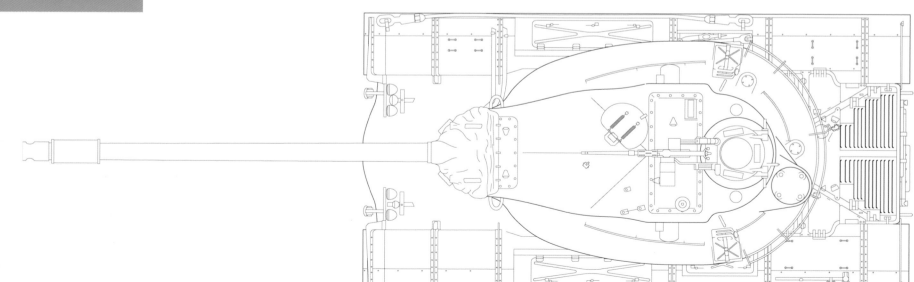

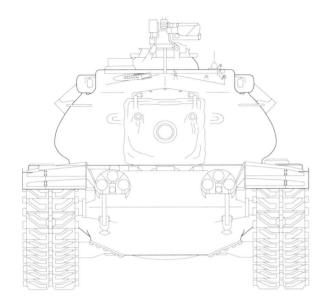
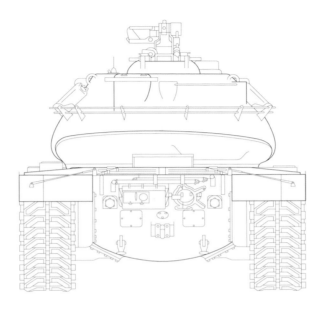

Left below: Power for the M103 was supplied by the Continental AV-1790 V-12, four-cycle, air-cooled engine. Various subtypes of this engine were used in the T43/M103 family of heavy tanks, producing varying values of horsepower and torque.
Right below: On top of the AV-1790 as viewed from the front left were two cooling fans for the engine. A carburetor with an inlet elbow was at the center of each side of the engine. At the bottom front of the engine is the engine-oil filter, to the rear of which is the generator. (TACOM LCMC History Office, both)

On display at Fort Hood, Texas, is this M103 heavy tank. It exhibits several attributes of an M103 that was never upgraded to M103A1 or –A2 standards, most notably the gunner's periscope cover in the rear position near the commander's cupola. In the background is an example of what in effect was the M103's successor as a heavy combat tank, the M1 Abrams main battle tank.

Top left: The mantlet of the M103 is observed from the front, allowing a view of the structure of the assembly without the dust cover, which is usually seen in vintage photos of the M103 heavy tanks. To each side is a coaxial machine gun port, with a cover over the right one. **Top right:** In the foreground on the 120mm gun barrel is the bore evacuator, which extracted toxic fumes from the bore of the gun after firing so the turret didn't become full of fumes. The headlight brush guards are the curved type introduced after the pilot T43s. **Above left:** The M103 at Fort Hood is viewed from the right side, showing the swell projecting to the front of the turret that incorporated the right trunnion bearing for the gun mount. Two lifting eyes are welded to the upper part of the mantlet to assist in removing the gun. **Above right:** In a right rear view of the M103, hand rails are on the turret, and to the rear of the bustle there is an upper rack for securing stowed equipment. Two projections in the bustle are for radio antenna mounts. At the rear of the turret roof is the commander's cupola.

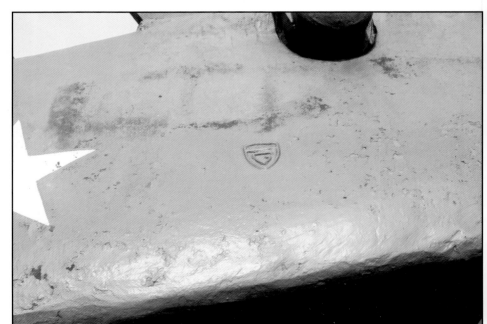

Top left: On the upper part of the rear of the hull are the two legs of the bipod-type travel lock for the 120mm gun. The yoke and latch that secured the gun barrel to the travel lock is to the right. The external telephone box, normally mounted on the left side, is missing. **Top right:** The curved contours of the side of the turret are apparent in this view from the left rear. The low rear deck that covered the engine compartment was a feature from the T43 to the M103A1; the many tanks converted to M103A2s had a significantly higher deck. **Above left:** On the rear deck of the M103 is a housing for the muffler, seen from the left rear; three of the elongated exhaust ports are visible here. To the side are grab handles for lifting the grille doors. Above and in the background is the underside of the turret bustle. **Above right:** On the left side of the turret below the housing for the left objective of the rangefinder on an M103 in Radcliff, Kentucky, is the trademark of the turret manufacturer, General Steel Castings Corporation: a union shield with a stylized letter G located in the center.

The T43E2/M103A1

Tests of the T43s and M103s indicated that these vehicles suffered from subpar sighting systems. To remedy this shortcoming, the Army modified two T43E1s, installing turret baskets and a new T52 stereoscopic rangefinder and T33 ballistic computer to be operated by the commander, not the gunner, and moving the gunner's station forward in the right side of the turret, providing him with a direct-sighting telescope and a repositioned gunner's periscope. Although the Army did not convert any more T43E2s, the Marines converted 219 of their T43E1s to T43E2s, such as this one, and these were standardized as the 120mm Gun Tank M103A1. (NARA)

While the first T43E2 was shipped to Aberdeen Proving Ground for testing in February 1956, the second prototype, shown here, was not shipped until June 1956. The second prototype was sent to Fort Knox. As the T43E2 differed from the T43E1 primarily in fire control, a considerable amount of the testing involved firing. During these tests, severe erosion of the blast deflector was noticed, and the component redesigned accordingly. (Patton Museum)

The Marines were not impressed with the Army's M103, and continued to press for resolution of all 114 deficiencies, which included further improved fire control and inclusion of a turret basket. In achieving these goals, an all-electric turret operating system was introduced, and the turret crew rearranged. The vastly improved tank was initially known as the T43E2, but on 17 May 1957 was type classified Standard as the M103A1. The Marines ultimately took delivery of 219 of the type, with the one vehicle shortfall being the result of the retention of one by Army Ordnance for testing. The USMC registration numbers assigned were 232954 through 233172.

Somewhat ironically, when the U.S. Army took heavy tanks to Europe in 1957 to counter the Soviet threat, the tanks they took were 72 borrowed USMC M103A1s.

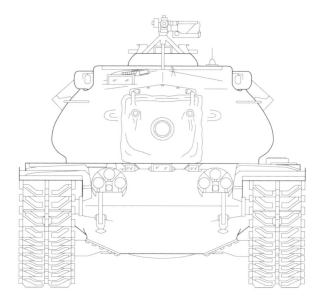
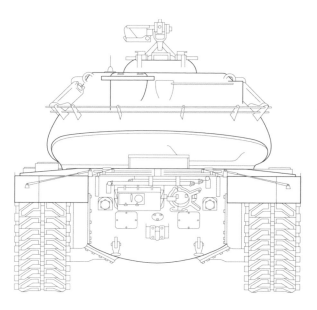

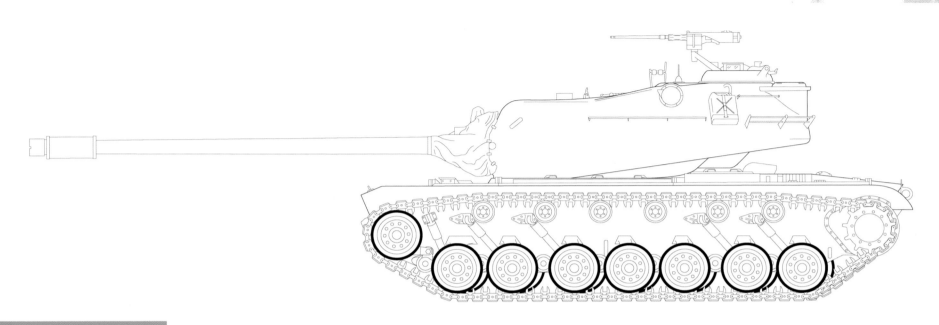

M103A1

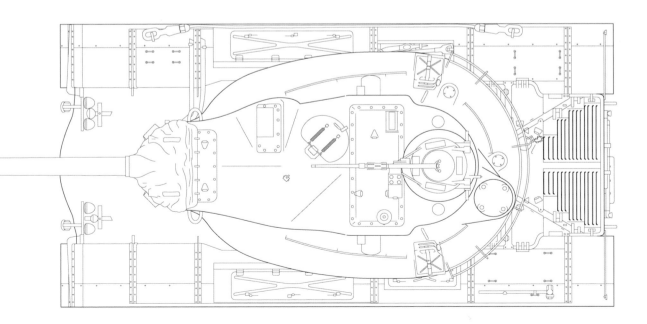

Top left: A key identifying feature of the T43E2/M103A1 compared to the M103 was the presence of the cover for the gunner's telescope to the front of the loader's hatch. The right coaxial machine gun was eliminated and a gunner's sighting telescope was installed in its place. **Top right:** The new location for the box-shaped gunner's periscope cover just forward of the loader's hatch is clearly visible in this photo of a T43E2 on a test range. Most of the other changes in the T43E2 were internal, involving improvements to the fire-control system. **Above left:** In an overhead view of a T43E2 on a test range, the position of the box-shaped enclosure for the gunner's sight on the top of the turret is clearly visible to the front of the loader's hatch. Radio antennas are on the mounts at the center rear and the right rear of the turret. **Above right:** Another change on the T43E2 and M103A1 was a new commander's cupola with a manually operated rather than remotely controlled .50-caliber machine gun mount. The electro-hydraulic elevating and traversing system was replaced by an amplidyne system. (Patton Museum, all)

The crew of a USMC M103A1 prepares to fire the 120mm gun during tests at Camp Pendleton, California, in April 1959. This was toward the end of the Marines' conversion of its T43E1s to M103A1s; the last of the conversions were completed in July 1959. A code that appears to be C34 is painted in large figures on the side of the turret. In addition to the normal locations of the two liquid containers, two extra containers are stowed on the turret bustle rack. (NARA)

An M103A1 of the 1st Tank Battalion, 1st Marine Division, operates along the shore at Camp Pendleton in late May 1959. This probably occurred during a test exercise of landing M103A1s from landing craft. The tires hanging on the turret reportedly were an identification device used to signify the tank's assigned unit in the exercise. The Marine Corps envisioned that the M103A1s would be used early on in amphibious landings to knock out enemy fortifications and tanks. (NARA)

Top left: Two Marines are preparing to load ammunition from the pile of crates in the foreground into an M103A1 during a firing exercise at Camp Pendleton in June 1959. With the tripod travel lock, it was possible to leave the unit in the raised position when not in use. **Top right:** An M103A1 of C Company, 1st Tank Battalion, 1st Marine Division, comes (PHIBLEX) Swan Dive on 15 February 1960. The code C-21 is marked on the bow and the turret. **Above:** Members of the 2nd Infantry Training Regiment move forward with M103A1 code Y52 at Camp Pendleton, 15 March 1960. In addition to the 1st Tank Battalion, USMC M103A1s were distributed to the 2nd and 3rd Tank Battalions and to armor

Marines of the 2nd Infantry Training Regiment work with the crew of an M103A1 to knock out a pillbox during exercises at Camp Pendleton on 15 March 1960. The Marine in the foreground is using the tank's external telephone to give directions to the crew. (NARA)

Due to the large size of the turret bustle, in order to do any work on the engine of the M103 series, the turret needed to be rotated past 90 degrees. Such is the case with this USMC M103A1 photographed at Camp Pendleton in the late sixties. One of the engine deck plates has been placed on top of the turret. This color photo gives a good impression of the USMC green seen on the M103 and the M48 to its left. Also seen to good advantage here it is the very long barrel length of the main gun. A cover has been installed over the muzzle brake. (Mike Kalbfleisch collection)

The commander of a Marine M103A1 observes the effect of a round just fired from the tank's 120mm gun on the distant hillside during a training exercise of Vieques, an island of Puerto Rico, on 25 April 1962. The tank was assigned to the 2nd Tank Battalion, 2nd Marine Division, based on the East Coast of the United States. The U.S. Navy and Marine Corps used the eastern part of the island as a firing and bombing range throughout the Cold War era. (NARA)

The M103A2

In 1961, the Marine Corps decided not to adopt the new M60 tank, opting instead to upgrade its M103A1s with certain M60 components. Three pilots were thus converted and designated M103A1E1. Despite some assertions that the Anniston Army Depot converted the Marines' fleet of M103A1s to this new standard, in fact the Red River Army Depot performed the conversions of 153 M103A1s starting in 1962 and another 52 in fiscal year 1968. The conversions were standardized as the M103A2 in December 1962, and the Marines began receiving them by spring 1964. Among the new features in this tank were the Continental AVDS-1790-2A diesel engine with improved range and torque, larger fuel cells, an M24 convergence-type rangefinder, and a redesigned engine cover with grilles at the rear of the hull. Here, an M103A2 rests on a tank transporter composed of an M123 tractor and M747 trailer. (Patton Museum)

The second of the two pilot M103A1E1s was photographed on a test range. The engine cover had a pronounced hump compared with that of preceding T43s and M103s. On the fenders to each side of the engine cover was a new, M60-type air-cleaner housing. The heat shield for the bottom of the turret is still installed here, but on production vehicles, it was eliminated. Interestingly, this tank mounts M60-style aluminum ribbed road wheels. (Don Moriarty collection)

Beginning in May 1961 the Marines began to study extending the service life of their M48 and M103A1 tanks. They opted to follow an Army-developed program for the M48, which included a coincidence range finder and significantly, the engine and engine compartment from the new Diesel M60 medium tank. On 5 June 1961 two tanks were ordered modified for testing, becoming M103A1E1.

In addition to the installation of the AVDS-1790-2A Diesel engine, the CD-850 transmission was upgraded to the -6 model. The reconfigured engine compartment allowed larger fuel tanks to be installed, increasing fuel capacity from 280 to 440 gallons. The change in powerplant resulted in a reshaped rear engine deck that now resembled that of the diesel powered M48A3 and M60. Series modification of the M103A1 to the new standard, which was designated M103A2, began in August 1963. The Marines converted 218 of their M103A1s into M103A2s (again, losing a single example to Army Ordnance as a test article) and used them up to 1972.

Although only 300 were produced, the M103 was by far the heaviest and most powerfully armed tank fielded by U.S. forces until the M1 Abrams arrived many years later.

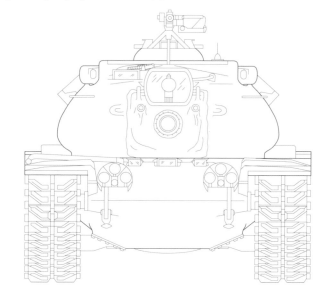
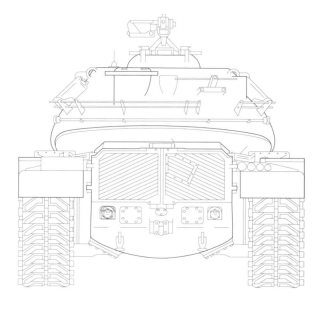

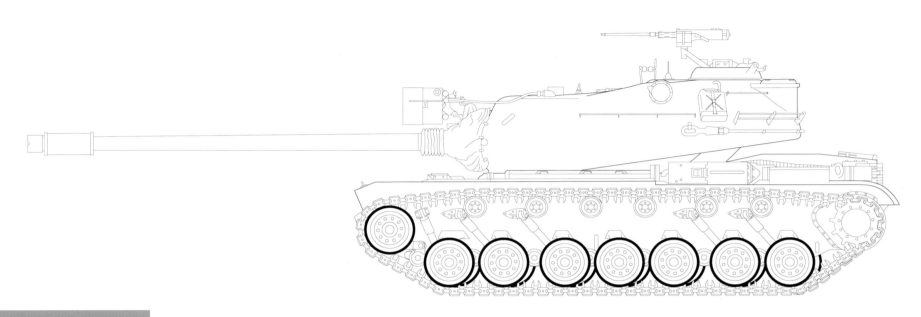

M103A2

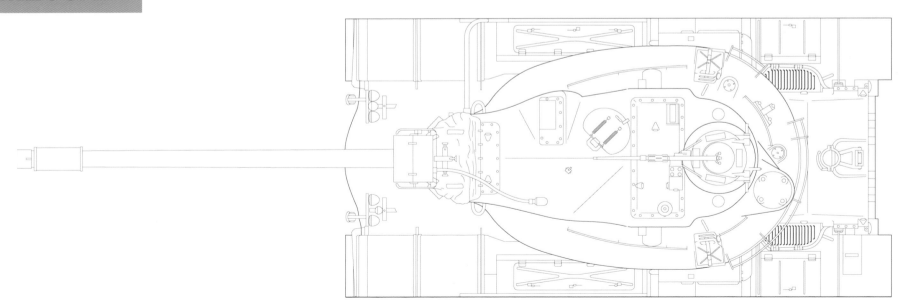

Top left: An M103A1E1 pilot is viewed from the rear, showing the new M60-style doors with grilles on the rear of the hull. Below the center of the doors was a tow pintle. Externally, the turret and armaments remained substantially the same as on the previous M103s. **Top right:** The raised or humped rear deck of the M103A1E1 pilot, as contrasted with the flatter rear decks of the M103 and M103A1, is evident in this rear view. A .50-caliber machine gun on a pintle mount, with an ammunition-box holder on the right side, is atop the commander's cupola. **Above left:** M103A1E1 pilot two is viewed from above, displaying details of the M60-type engine cover and grilles. Gone was the bulky, tall travel lock, replaced now by the same type of swiveling travel lock as carried on the M60, located on the center rear of the cover. **Above right:** The turret of M103A1E1 pilot two is now traversed to the rear and the 120mm main gun is secured in the travel lock. The new air-cleaner housings are on the fenders to each side of the gun mantlet. The mantlet dust cover now had an accordion-type front extension. (Patton Museum, all)

Top: This USMC M103A2 was photographed at Aberdeen Proving Ground on 4 October 1962. "USMC" and the registration are faintly visible in yellow paint on the side of the M60-type stowage box on the rear of the fender. On the mantlet is a searchlight mount. (NARA) **Above left:** Another USMC M103A2 is displayed. The main gun was the 120mm Gun M58 in the Mount M89A1. The turret had a maximum traverse rate of 17 seconds for a full 360 degrees. The gun had a maximum firing rate of five rounds per minute. (TACOM LCMC History Office) **Above right:** Registration number 233103 is marked on the side of the left air-cleaner housing on this USMC M103A2 in late 1966. The accordion-type front extension to the mantlet dust cover appears to good advantage. A foundry trademark is below the turret handrail. (Patton Museum)

M103A2 USMC registration number 233139 is the subject of this June 1964 color photo. There was variation in where the USMC registration number was placed on M103A2s; here it is in light yellow on the bow and on the side of the stowage box on the fender to the front of the air cleaner. The mantlet cover was of a different color than the green of the tank, being olive drab or an approximation of it, with a semi-gloss sheen. (NARA)

Left: An M103A2 assigned to the 1st Tank Battalion, 1st Marine Division, advances against "enemy" forces during Operation Silver Lance at Camp Pendleton, California, on 3 March 1965. Several tires are stacked on top of the front of the turret, presumably as visual symbol of the tank's unit during the maneuvers. A marking, apparently a tactical symbol, containing a triangle within a circle is on the bow. **Top right:** During Operation Silver Lance at Camp Pendleton on 1 March 1965, a Marine M103A2 appears in an aerial photograph. On the rear deck of the tank is ample local camouflage in the form of tree branches. Tires are stacked on the front of the turret roof. **Above right:** Two M103A2s assigned to the Aggressor Forces form up to combat the Marine Force during Operation Silver Lance, 9 March 1965. The tank in the foreground has a searchlight, and the farther tank has a sign with a triangle inside a circle on the turret. (NARA, all)

A day after the preceding photo was taken, an M103A2 advances during Operation Silver Lance. The ubiquitous tires on the turret roof must have made forward vision difficult for the commander and the gunner. A five-gallon liquid container is on the rear of the hull. (NARA)

In another 10 March 1965 photograph from Operation Silver Lance, an M103A2 stirs up considerable dust as it proceeds down a trail. This operation was conducted by III Marine Expeditionary Force to test new procedures for landing and deploying armored forces. (NARA)

Two M103A2s assigned to D Company, 5th Marine Tank Battalion, are engaged in training maneuvers at Camp Pendleton on 29 May 1967. The closer tank has registration number 233034 and a crest painted on the side of the air cleaner housing—probably the 5th Marine Tank Battalion insignia. (NARA)

A column of M103A2s assigned to D Company, 5th Marine Tank Battalion, conduct firing practice on a range at Camp Pendleton on 6 June 1967. The code on the turret of the first tank, D25, stood for the fifth tank of the 2nd Platoon of D Company. (NARA)

The fourth M103A2 of 2nd Platoon, D Company, 5th Marine Tank Battalion, has the 120mm gun traversed to the right during training at Camp Pendleton, California, on 6 June 1967. A light-colored crest is visible on the air cleaner, the liquid containers, and the left rear fender was present on M103A2s observed at Camp Pendleton during this period, and it also was marked on the upper center of the bow. (NARA)

Top left: Turrets pointed aft, a column of USMC M103A2s proceeds down a trail in a wooded area. A close examination of the lead tank reveals the stowage boxes and the air-cleaner housing toward the rear of the fender: identification characteristics of the M103A2. **Top right:** The driver of an M103A2 is in the hatch-open position, offering far greater visibility than when he is buttoned-up in the tank. Above him is the turret bustle, angled to provide the driver with headroom. Two five-gallon liquid containers are piggybacked on the turret. **Above left:** The commander of an M103A2 surveys the situation during a training exercise. Instead of the usual .50-caliber machine gun, a .30-caliber machine gun is on the cupola mount. One of the stowed five-gallon liquid containers is marked as a water container.

Above right:
Smoke obscures the background as the crew of an M103A2 conducts firing practice. Three spent propellant casings are lying to the ground to the side. The ammunition of the 120mm Gun M58 comprised a separate projectile and propellant cartridge. (Patton Museum, all)

Crews of a line of four M103A2s take a break during firing practice. The closest tank bears USMC registration number 232987 on the air-cleaner housing. The 120mm Gun M58 of these tanks had a muzzle velocity of up to 3,750 feet per second. The maximum range varied by type of ammunition, with, for example, a range of 25,290 yards with high-explosive anti-tank tracer (HEAT-T) or armor-piercing shot tracer (AP-T) projectiles. (Patton Museum)

A line of USMC M103A2s is in storage at the Yermo Annex, Marine Corps Supply Center, Barstow, California, on 17 September 1975. The Marine Corps phased out the M103A2 from active service in 1972, after which the Corps adopted the M60A1 tank. Approximately 25 M103s are thought to survive, in varying states of preservation and mostly in display-type conditions. Note that the bore evacuators and the canvas mantlet covers have been removed from these vehicles. (NARA)

This M103A2 is on display at the Marine Corps Mechanized Museum at Camp Pendleton, California. The name "GRAVE DIGGER" is painted in yellow on the bore evacuator, and the unit/vehicle code D30 is marked on the side of the turret. (Chris Hughes)

Top left: The Jacques Littlefield Collection preserves this M103A2. Mounted on the top of the mantlet is a xenon searchlight. On the front of the hull are two lifting eyes with swiveling tow shackles on the bottoms. Also visible from this angle are the driver's periscopes. **Top right:** The left headlight array includes a service headlight and a blackout headlight, between and below which is a blackout marker lamp. To the right, details of the front support bracket for the fender, the rubber, chevron track blocks and track connectors are in view. **Above left:** In a view facing aft under the front left corner of the hull, to the right are the inner side of the dual compensating (idler) wheel and the compensating-wheel link. Jutting from the hull at the bottom of the photo is the support housing for the front suspension arm. **Above right:** On the right headlight array are a blackout service light on the inboard side, a service headlight on the outboard side, and a blackout marker lamp below and between the service lights. Missing above the lights is the horn. To the left is the front fender support.

Top left: Standard for the M103A2 were the T107 center-guide tracks, a double-pin type with rubber chevron blocks. The tracks were 28 inches wide, with a pitch of 7.09 inches. On the inboard and the outboard side of the track blocks are track connectors. **Top right:** The right dual compensator wheel is viewed from the inside. Adjusting the compensating wheel to the front or to the rear achieved the proper tension of the track. To the lower right are the pin and the cotter pin that secure the right front tow shackle in place. **Above left:** The right dual compensating wheel and the two front right dual road wheels are shown. Between the compensating wheel and the front road wheel is the compensating-wheel link. The front three road-wheel arms are linked to the shock absorbers visible here. **Above right:** A close-up view over the top of the front right dual road wheel shows the two forward shock absorbers and the compensating-wheel link. The tops of the shock absorbers are pinned to heavy-duty shock-absorber brackets that are bolted to the side of the hull.

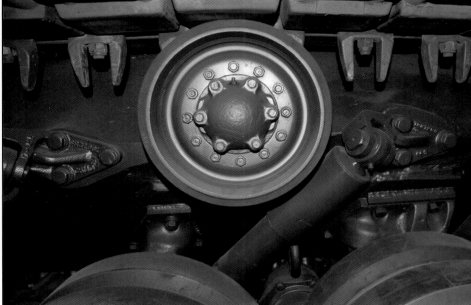

Top left: The dual road wheels are secured to the hubs with ten hex lug nuts, and the hub caps are fastened with six hex screws, and a grease fitting is at the center of each of the hubs. The tires were size 26-6. The Goodyear logo is visible on the edges of these tires. **Top right:** There were six track-return rollers on each side, and they were of dual-wheel construction with rubber tires, 12 hex fasteners, and a hubcap with six hex fasteners and a grease fitting on the side. A view also is offered of the dual-horned track guides. **Above left:** The engine cover of the M103A2 was taller than that of preceding M103 types to accommodate the larger engine. Whereas preceding types had the engine exhaust on the rear deck, the M103A2 grilles served to vent exhaust gasses and cooling air. **Above right:** Two technicians inspect a Continental AVDS-1790-2A Diesel engine of the type that powered the M103A2. At the top right is a turbo supercharger connected to the exhaust line. The 1791.7 cubic-inch engine produced 750 gross horsepower at 2400 rpm. (US Army Quartermaster Museum)

The two rear doors of the hull had grilles for venting exhaust and cooling air. The right door had a fitting for a deep-water fording kit. Below the doors are two transmission-access plates, between which is a tow pintle. Above the doors is the travel lock. (Chris Hughes)

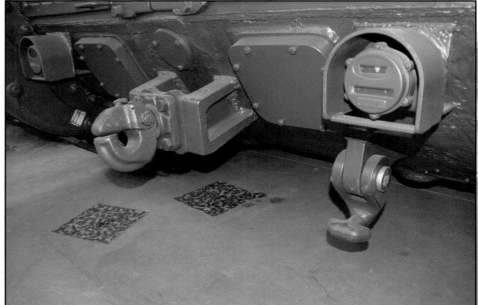

Top left: The rear of the right fender of an M103A2 is shown, including a clear view of the chevron pattern on the rubber track blocks. The rear fender support is triangular with lightning holes that taper in size. To the front of the support is a stowage box. (Chris Hughes)
Top right: The right tail-light assembly (foreground) and the left one (background) on the M103A2 were of different design, the right one having a blackout stoplight over a blackout marker lamp, and the left one with a tail light/service stoplight over a blackout marker lamp. **Above left:** On the each side of the bottom of the vertical rear of the hull is a round plate for accessing the brake rods. Toward the right is the plate on the right side. The bulge in the hull below the plate is the location of the right transmission oil-drain cover. **Above right:** In a view of the lower right of the rear of the hull of an M103A2, to the right is the right final drive, at the center is the right rear tow shackle, above which is the right transmission access cover. Casting numbers are present to the right of the tow shackle.

Top left: The tow pintle is mounted on a heavy-duty steel bracket welded to the rear of the hull. The small round plate above the bracket is the transmission center access plate. At the top of the photo are the grab handles at the bottoms of the rear hull door/grilles. **Top right:** The tilted pentagonal frame on the upper inboard corner of the right rear door provided the mounting surface for the stack of a deep-water fording kit. The stack had an elbow at the bottom that mounted on this frame using the 11 hex screws visible here. (Chris Hughes) **Above left:** Whereas the external telephone box from the T43 to the M103A1 heavy tanks was on the rear of the hull, the external telephone box on the M103A2 was on the rear of the left fender, as seen here. The door for accessing the phone is hinged at the bottom. **Above right:** The external telephone box door is open, showing the handset telephone, coiled cord, receptacle, and interphone control box. To the lower right is the left tail light assembly, featuring a red, oval stoplight element at the top and a blackout marker at the bottom. (Chris Hughes)

Top left: The interior of the external telephone box on the rear of the left fender is shown in closer detail. The telephone handset is held in a cradle on the door. Also on the door is a container for the telephone cord. To the right is the intercom control box. **Top right:** The intercom control box in the external telephone box of the M103A2 has controls for volume, intercom or radio selection, and line. The external telephone box allowed infantrymen outside the tank a ready means to communicate with the crew. **Above left:** A cable and jack in the upper left corner of the external telephone box are depicted. Mounted on top of the box is a small red signal light. (Chris Hughes, three) **Above right:** The manner in which the external telephone box, shown with the door closed, is attached to the rear of the left fender is displayed. To the front of the box is a stowage box, followed by the left air-cleaner housing. Tow-cable clips are on the rear of the turret.

Between the telephone box and the stowage box is a lifting ring. The travel lock for the 120mm gun is missing from the slightly recessed spot at the center rear of the rear deck. Features of the air-cleaner housing, the rear of the turret and of the commander's cupola are in view. A stowage rack comprising two curved steel rods attached to struts is on the lower part of the bustle. Below the liquid container is a clamp for the tow cable. (Chris Hughes)

Top left: The stowage box on the rear of the right fender is viewed from above. Inboard of the box is a grille/door, with two grab handles in view. To the rear of the grille/door is the right rear lifting eye, to the rear of which is the rear fender support with lightening holes.
Top right: The rear deck of the M103A2 is observed from the right, with the rear of the turret in the foreground. At the center rear of the deck is an indentation where the travel lock for the main gun was mounted. The engine-compartment door and grab handle are at the center. **Above left:** A travel lock on an M103A2 is seen close-up from the right side, with the front of another armored vehicle in the upper left background. The turret is traversed to the rear, and the 120mm gun is clamped into the lock. When not in use, the clamp swung down. **Above right:** In a view of an M103A2 from behind the left rear of the hull, the raised travel lock is at the top of the photo. The guards for the tail-light assemblies consist of a hood shaped like an inclined letter D, with a metal bar across the bottom of the guard.

Top left: At the rear of each side of the M103A2 is a sprocket assembly, consisting of an outer sprocket and an inner sprocket, each of which is mounted with 11 hex screws onto a drum with oval slots in it. Each sprocket has 11 teeth. This is the left sprocket. **Top right:** The engine cover and grilles, the external telephone box, the left rear stowage box, part of the left air-cleaner housing, and the left rear of the turret are in this view. The stowage box has a single operating handle. Footman loops are on the turret for securing gear. (Chris Hughes) **Above left:** Taken from the engine deck to the rear of the turret, facing to the left, in this view are, left to right, the external telephone box; the left rear fender support; the left rear lifting ring; and stowage box. Note the coupling for the telephone line on the fender support. **Above right:** Elements on the forward left part of the M103A2 visible here include the driver's open hatch and periscopes, pioneer tools, the fender and its supports, and a long stowage box. On the number 4 on the turret is the trademark of the General Steel Castings Corporation. (Chris Hughes)

Top left: On the opposite side of the turret from the preceding photo, the right side of the turret and the handrail is to the left, with the long stowage box and its three operating handles to the right. **Top right:** The front of the left stowage box on the fender adjacent to the turret is viewed from the front. The operating handles of the hinged cover of the box have holes near the pivoting ends that match holes in lugs welded to the lid, to allow the attachment of padlocks. **Above left:** Facing from the left fender (bottom) to the left side of the bow, a spare track link, a shovel, and a sledgehammer are secured in place. Next to the head of the sledgehammer is the housing for the exterior controls for the fixed fire extinguisher. **Above right:** The driver's station was equipped with three fixed periscopes through the hull and one periscope on a rotating mount on the hatch cover. Inside the far corner of the hatch is the operating mechanism for the hatch cover. To the lower right is the heater exhaust outlet. (Chris Hughes, both)

Most of the left side of the turret of the M103A2 heavy tank is in view from this angle. The commander's .50-caliber machine gun mount appears above the housing for the left objective of the rangefinder. An angled lifting eye is on the side of the turret toward the front. A xenon searchlight is mounted above the 120mm gun mantlet. Next to the track in the center foreground is the left front lifting eye and tow shackle.

Top left: The small, indented circle on the forward extension of the side of the turret marks the location, inside the turret, of the left trunnion of the gun mount. Above and below that spot are block-shaped attachment points for the dust cover of the mantlet. **Top right:** From left to right on the turret roof are the cable and receptacle for powering the xenon searchlight; the loader's hatch with the cover open; the removable cover for the rangefinder; the commander's cupola and machine gun mount; and the turret ventilator. (Chris Hughes) **Above left:** Mountings were provided on the upper part of the mantlet of the M103A2 for a xenon searchlight, typically an AN/VSS-1 model designed for mounting on tanks. The searchlight is viewed from the left rear, showing the power cable and side handles. **Above right:** The power cable for the xenon searchlight is plugged into a receptacle on the turret roof. To the left is the cover for the gunner's periscope, to the rear of which is the loader's hatch. Casting numbers are visible on the roof of the turret toward the right of the photo.

Top left: The commander's cupola near the rear of the turret roof of the M103A2 incorporated four M17 periscopes for viewing outside when he was in the tank. The front periscope is between the legs of the gun mount. The commander also operated the M24 rangefinder.
Top right: The dome-shaped housing for the left objective of the rangefinder is viewed from the front. A similar housing is on the right side of the turret. A lifting ring is on the top of the housing. In the background is a five-gallon liquid container secured to its holder. **Above left:** The left housing of the objective of the M24 coincidence-type rangefinder is viewed from the side. On the roof of the turret between the two objective housings is the removable rangefinder cover, fitted with triangular lifting eyes and a small ventilator hood. (Chris Hughes) **Above right:** The commander's M11 cupola is viewed from the left side, showing the two periscopes and their guards on the side of the cupola ring. The rear-facing periscope for the cupola was mounted on the cupola hatch cover. (Chris Hughes)

A handrail extends for several feet to the front of the liquid container. Below the liquid container is a clamp for the tow cable. Around the bustle of the turret is a rack for combat packs consisting of upper rails, dual bottom rails, and several footman loops. (Chris Hughes)

Top left: Two steel castings are welded to the turret bustle to hold radio antennas. This is the right one, to the rear of the right liquid container. The part number is visible on the casting. At the top of the casting is the antenna support base, which holds the antenna when installed. **Top right:** Below the top rail at the center rear of the turret bustle is welded a bracket for storing the xenon searchlight when not installed on the mantlet. To the left is the casting for holding the left antenna and antenna support base. Varying textures of the armor are apparent. **Above left:** The turret ventilator is at the left rear of the bustle. Seen here is the ventilator cap or hood. Partially surrounding it is a splash guard. Four holes are on top of the cap for fastening it in place. To the bottom right is the brown insulator of the antenna base. **Above right:** The left antenna is located at approximately the front-rear centerline of the turret bustle. The base is attached with four hex screws with locking nuts. Radios specified for use in the M103A2 heavy tank were the AN/GRC-3 to -8 and the AN/ARC-3 or -27.

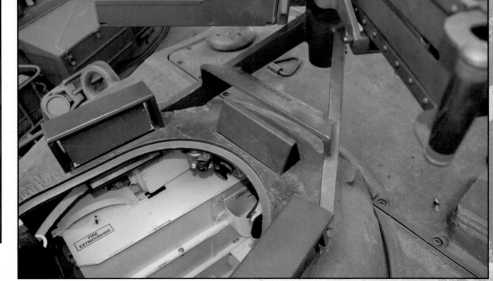

Top left: The loader's hatch cover, viewed from the right side of the turret, has a padded head guard and a lock-bolt mechanism on the inner side. The thickness of the roof armor is visible. To the right is the rear of the gunner's M29 periscope and its armored guard. **Top right:** On each the outboard side of each hatch-cover hinge on the cupola is a can-shaped housing containing a spring that acted as a counterbalance to ease the process of opening and closing the heavy, armored hatch cover. The right one is visible from this angle. **Above left:** is dished out to accommodate the commander's head. A rear-facing periscope is mounted in the rear of the hatch cover, secured with toggle bolts and wing nuts. **Above right:** The legs and the crossbar of the machine gun mount formed a guard for the front periscope of the cupola. Inside the commander's hatch is visible

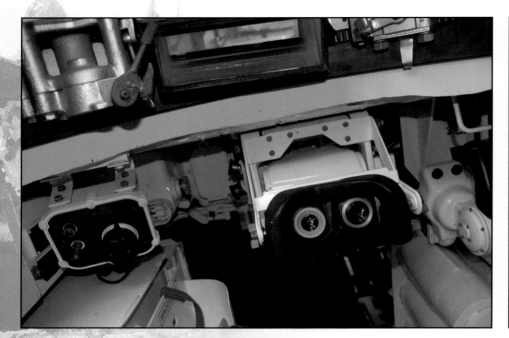
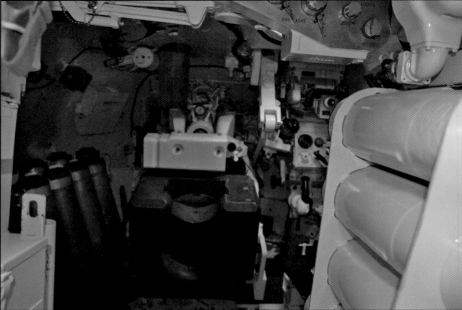

Top left: Looking into the commander's hatch to the front, at the top on the inside of the cupola are, left to right, the cupola brake, the forward periscope of the cupola, and the cupola lock. Below are the rangefinder eyepieces and the commander's turret-control stick. **Top right:** This is a view from the commander's seat in looking forward in the turret. Propellant-cartridge racks are on both sides in the foreground. In the center background is the breech of the 120mm Gun M58. To the right of the breech is the gunner's station. **Above left:** Inside the rear of the turret of the M103A2 facing left and forward from the stowage points for a canteen and a flashlight. To the right is radio transmitter and receiver equipment. **Above right:** From the commander's seat looking to the front right, to the upper right is the commander's stick that enabled him to control traverse and elevation, overriding

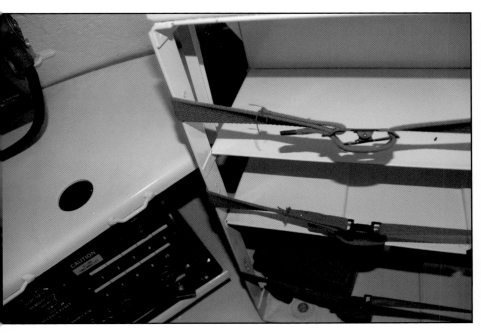

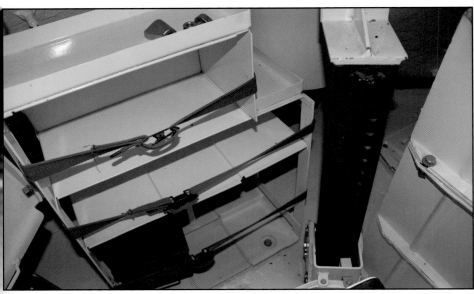

Top left: To the rear of the commander's seat in the turret, to the left is radio equipment. A sticker on it cautions that it weighs 61 pounds and is "two-man lift." To the right are racks for storing .50-caliber ammunition, with webbing retainer straps with adjustable buckles. **Above left:** Racks for .50-caliber ammunition in the rear of the turret are shown more completely. At the top of the racks is an oddments tray. To the right is the support stanchion for the commander's seat, which allowed the seat to be locked at different heights.

Right: The top of the turret and the rear of the xenon searchlight are observed from the loader's hatch. The searchlight is attached to the mantlet by means of a three-point ball-and-socket mount. Two latches on top of the unit allowed the cowling to be opened for servicing.

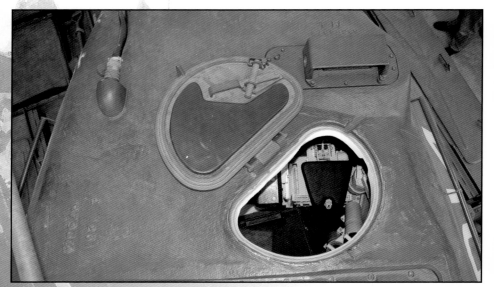

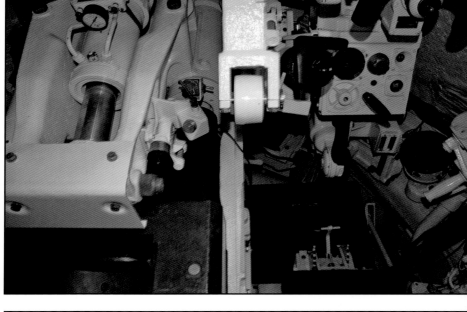

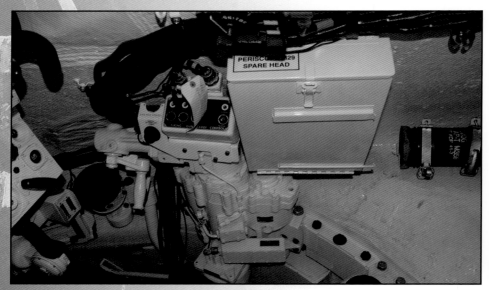

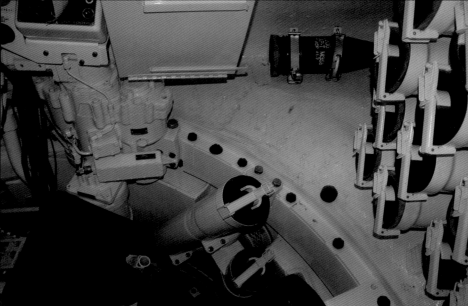

Top left: The loader's hatch and its open cover are viewed. Inside the turret below is the gunner's seat. To the right of the hatch cover is the gunner's periscope and its guard. Casting numbers are visible on the turret surface to the lower left and to the front of the periscope. **Top right:** Looking into the turret through the loader's hatch, the top of the gun breech is to the lower left; above it are the equilibrator and the equilibrator bracket. To the right are the gunner's periscope, the telescope, control handle, and ballistic computer. **Above left:** To the gunner's right, the dark, round object is the azimuth indicator. At the center of the photo is the traverse gear, above which, left to right, are the manual traversing handle; gun-selector switch box; and Periscope M29 spare head storage box. **Above right:** To the rear of the gunner's seat, the back of which has been removed, are two vertical stowage tubes for 120mm propellant cartridges. To the far right are propellant tubes to the right of the commander's station. A 120mm AP-T projectile is in clamps at the top.

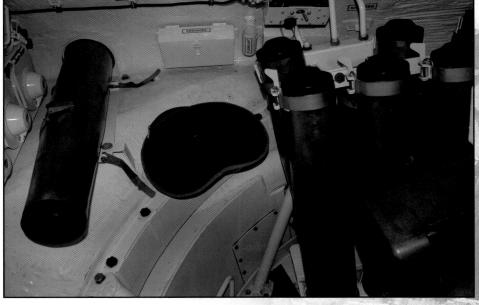

Top left: Facing the left rear of the turret, to the right are the left loader's seat and a grenade box. At the center is a propellant cartridge on a ready rack. At the bottom left is a perforated, hinged guard that was raised to protect the commander's legs from ejected cartridges. **Top right:** From behind the gun breech, lower right, facing the left side of the turret, propellant cartridges are stowed vertically with clamps holding the tops. At the top are, left to right, the loader's back cushion and a box with a ventilator switch and an outlet. **Above left:** In a continuation of the preceding view, above the ventilator switch box are the loader's gun-safety switch box and, at the top, the loader's intercom control box. At the upper center are brackets for storing the loader's seat. To the right is the equilibrator. **Above right:** Several types of 120mm ammunition were carried in the M103-family heavy tanks. Left to right, they are: HE-T T15E3; T21E1 propellant; WP-T T16E3; AP-T T116E6; T30E1 propellant; and TP-T T147E3.

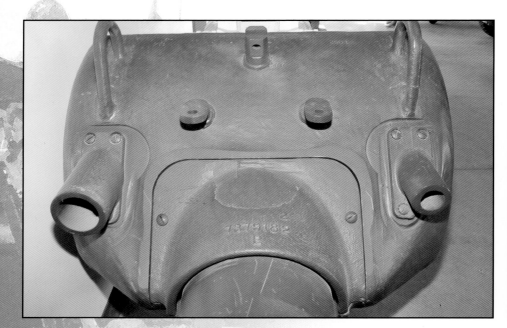
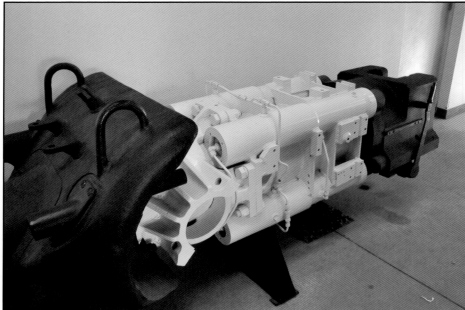
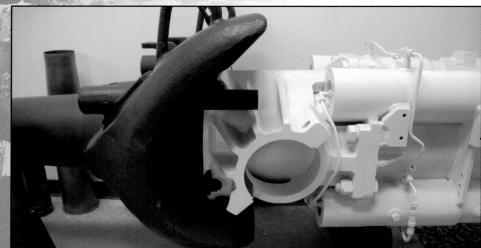
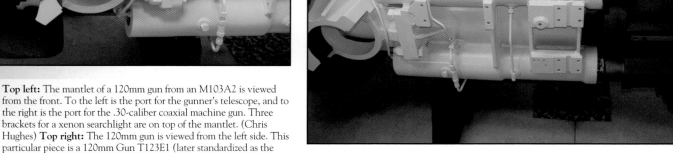

Top left: The mantlet of a 120mm gun from an M103A2 is viewed from the front. To the left is the port for the gunner's telescope, and to the right is the port for the .30-caliber coaxial machine gun. Three brackets for a xenon searchlight are on top of the mantlet. (Chris Hughes) **Top right:** The 120mm gun is viewed from the left side. This particular piece is a 120mm Gun T123E1 (later standardized as the M58) on a Mount, Combination, Gun, M89A1. The white structure is the cradle, which also incorporates two recoil cylinders per side. (Chris Hughes) **Above left:** To the front of each side of the cradle to the immediate rear of the mantlet is a trunnion; this is the left one. The coaxial machine gun and its cradle were on this side of the 120mm gun mount, and the tube for the machine gun barrel is visible on the mantlet. **Above right:** The left side of the 120mm Gun T123E1 on the Mount, Combination, Gun, M89A1 from the rear of the mantlet to the rear of the breech is in view. The equilibrator bracket is not installed above the gun. The breech-operating handle is on the side of the breech.

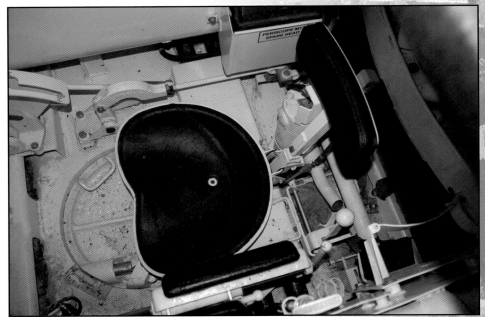

Left: As viewed from the rear, the breech features a falling breechblock, shown partially closed. Beveled corners are evident on the breech. The gun's nomenclature, ordnance number, and other data are stamped on the top of the breech. On the rear of the gun cradle is a nomenclature plate for the gun mount, including the serial number and a list of technical manuals pertaining to this type of gun mount. **Top right:** The driver's compartment is viewed through the driver's hatch, the front of which is at the bottom. Directly below is the seat. Below the seat in the floor of the hull is the driver's escape hatch. To the left, inside the hull, is the hatch-cover operating mechanism.
Above right: The driver's compartment is viewed with the right side of the compartment to the top. At the bottom is the driver's armrest. To the driver's right are the transmission shifter plate and part of the parking-brake mechanism. To the top right is a periscope-head box. (Chris Hughes, all)

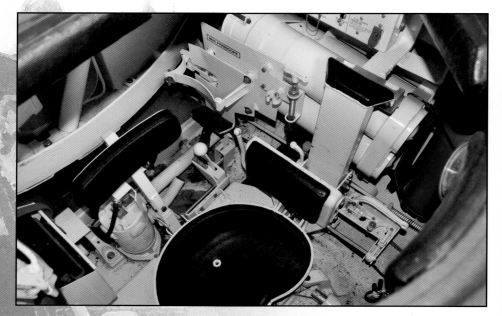
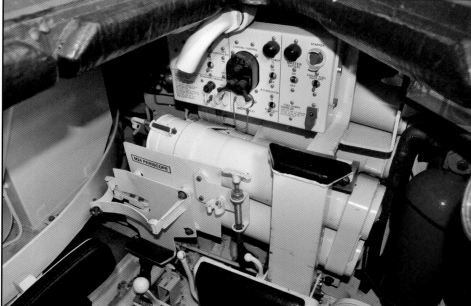

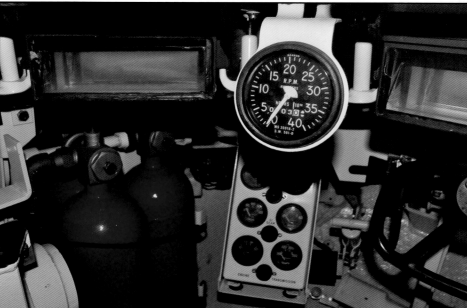

Top left: Looking through the driver's hatch to the left side of the driver's compartment, the seat-support column is next to the armrest. The seat could be raised and lowered on the column. Powder-cartridge stowage tubes are on the outboard side of the column. **Top right:** Above the seat column on the left side of the driver's compartment is the switch panel, including the master light switch at the center. A padded cushion is at the top of the seat column, next to which is a fuel-valve T-handle. To the far left is the turret basket floor. **Above left:** In the front left corner of the driver's compartment are two fixed, ated from within the tank or by using activating controls on the upper left part of the glacis. **Above right:** To the upper left of the steering wheel (right) is a tachometer with a timer. This instrument appears here between left and the center M27 periscopes of the driver's

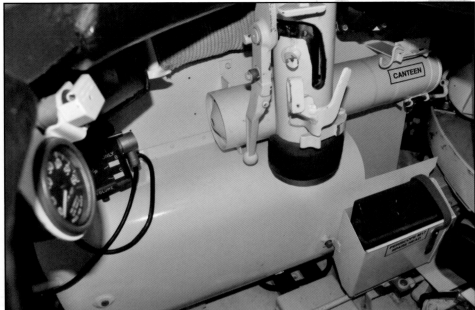
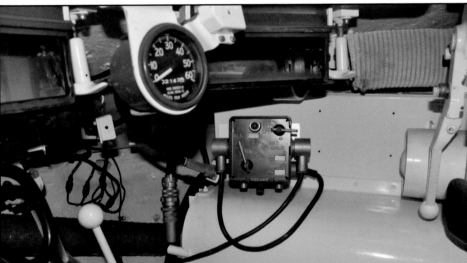
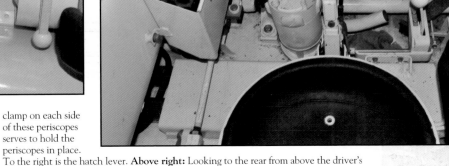

Top left: The tachometer and the speedometer flank the front periscope (top center). Below the speedometer is the transmission shift lever, and below the periscope is the steering wheel. The instrument panel changed in location and design from model to model in the M103. **Top right:** This view was taken from the front center of the driver's compartment looking to the right rear. To the top center is the operating mechanism of the driver's hatch cover. In the bottom half of the photo is a sheet-metal enclosure for ammunition stowage. **Above left:** At the center of the photo, on the right side of the driver's compartment, is the driver's intercom control box. Above it is the right periscope. A white spring clamp on each side of these periscopes serves to hold the periscopes in place. To the right is the hatch lever. **Above right:** Looking to the rear from above the driver's seat (bottom), under the floor behind the seat was a bilge pump and the batteries. To the rear is the turret basket. Before the M103A1, the T43s and M103s lacked a turret basket, making the loaders' tasks more difficult. (Chris Hughes, all)

M51 Tank Retriever

The Recovery Vehicle, Heavy, T51, standardized as the M51, was based on elements of the M103 heavy tank and was developed out of a need to field larger and more powerful recovery vehicles to handle the increasingly heavy tanks of the early 1950s. Work began on the T51 in early 1951, and following the production and testing of two pilots, it was released for production by Chrysler and standardized as the M51 in 1953. Shown here is T51 pilot number 3, the production pilot, on 10 February 1954. (Patton Museum)

Recovering combat vehicles, especially under hostile conditions, required specialized equipment. The considerable size foreseen for the T43 precluded the use of the military's then standard Sherman-based M32 and M74 retrievers. Hence, development of a new, larger retriever utilizing many T43/M103 components was initiated in February 1951. The first pilot of the new type, designated T51, began testing at Aberdeen Proving Ground in April 1953. Produced by Chrysler at the Detroit Tank Arsenal, the vehicle was largely satisfactory, and production was initiated soon thereafter. The most visible differences between the two test articles and the production models was the deletion of the initial rear outriggers and replacing them with a blade, similar to the one already in place on the front of the vehicle.

The M51 had a hydraulically driven 45-ton capacity recovery winch as well as a five-ton capacity auxiliary winch. Hydraulically actuated anchor blades located on both the front and rear of the M51 stabilized the vehicle during recovery and lifting operations. A crew of four operated the vehicle.

During 1954-1955 Chrysler built 187 of the vehicles, which had since been standardized as the M51. The M51 was used by both the Army and Marines. Notably, unlike the M103, the Marines did take the M51 to Vietnam. Unlike the heavy tank, the M51 was never upgraded from gasoline to Diesel engine.

Below left: A T51 pilot is viewed from the right rear. At the center of the hull was a crane rated at a maximum capacity of 60,000 pounds. On each side of the rear of the hull was a swiveling jack with a large, round, base plate for stabilizing the rear of the vehicle when lifting. **Right:** A T51 crosses a muddy slough. A ladder was provided on each side of the vehicle for access to the cab. The engine and transmission were located in the rear of the hull. The horizontal sheaves seen in the preceding photo have been replaced by two steel rods. (Patton Museum) **Below right:** The production Recovery Vehicle, Heavy, M51 varied in certain aspects from the T51. The M51 retained the front spade but substituted a hydraulically operated rear spade for the two stabilizing jacks. A cupola-mounted .50-caliber machine gun was added. (Patton Museum, all)

Top left: The crane is shown in its retracted position. It could be extended four more feet to the rear but at the expense of sacrificing half of the maximum 60,000 pounds of lifting capacity. A tow bar was stowed above the rear half of the fender on each side. **Top right:** Housed in the base of the crane were a crane winch for raising and lowering the boom and a hook winch for operating the hook cable. The crane could be swiveled 30 degrees to each side of center by means of a double-action hydraulic cylinder. **Above left:** The crane of an M51 is raised and traversed left. The rear spade, shown in the travel position, and the front spade were operated by cables from the and 7.09-inch pitch. **Above right:** Three crewmen are at the alert on an M51. The four-man crew rode in the cab during transit; they included the commander, driver, crane operator, and rigger. On the front of the vehicle is the front spade, used for stabilizing the vehicle and

Top left: An M51 assigned to B Company, 2nd Tank Battalion, 2nd Marine Division, tows a Swedish-made Stridsvagn L-60L light tank during the U.S. occupation of the Dominican Republic in 1965. A yellow acetylene bottle is secured to the rear of the cab. **Top right:** Marine M51 registration number 229146 of the 3rd Tank Battalion is being positioned on a landing craft at Dong Ha on 29 June 1967. A tow bar is attached to the front of the hull and is supported by a cable routed through the auxiliary-winch door on the cab front. **Above left:** An M51 stands by in the foreground next to a tank transporter loaded with an M48 tank at the 1st Force Service Regiment base, Da Nang, Republic of Vietnam in December 1967. This photograph shows the boom of the crane in its extended position. **Above right:** A Recovery Vehicle, Heavy, M51 is viewed from the rear as it churns through thick mud at the USMC base at Con Thien, Republic of Vietnam, on 16 December 1967. The vehicle is pulling an M50 Ontos self-propelled multiple 106mm recoilless rifle from the mud, visible to the right front of the M51. (NARA)

Repairs are underway on an M51 at the USMC base at Dong Ha, Republic of Vietnam, in 1968. The left sprocket assembly has been removed and is lying on the ground next to the man to the right. Black-and-yellow caution markings are painted on the crane hook. (NARA)

Based on the M103 heavy tank, the Recovery Vehicle, Heavy, M51, was conceived and built to recover the increasingly heavy tanks that were being introduced from the early 1950s. It featured a heavy-duty, extendable and traversable crane, front and rear stabilizing spades, and an array of winches. This surviving example of an M51 is painted overall in Olive Drab and is in an excellent state of preservation.

Top left: In a frontal view of the M51, the front spade is in the raised position. Above it on the right side of the glacis is a square armored door, which, when opened, allowed a line to be passed out from the auxiliary winch. Below the spade is the main-winch front cover. **Top right:** The auxiliary-winch door is observed close-up, with the spade at the bottom of the photo. Above the door is a cable anchor and, at the top, a cable roller. The large plate bolted to the glacis and fitted with lifting eyes was removable to provide access to the main winch. **Above left:** The auxil- The door is hinged on the inboard side, and on the outboard side are two swiveling dogs for securing the door shut. The dogs are in the open position in this photo. **Above right:** The headlight assemblies were mounted to each side of the front of the cab and were protected by brush guards made of welded steel slats. This one is on the right side. Alongside the brush

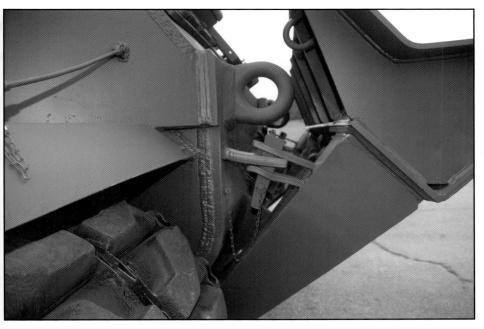
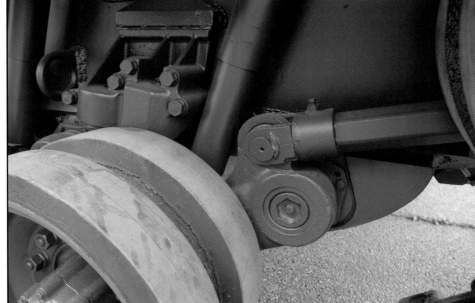

Top left: A view of the front spade from the right side gives a good idea of the thickness of the steel of that assembly. Weld seams on the bodywork are apparent. Behind the spade is the front right lifting eye. To the upper right is a support rod for the front end of the fender. **Top right:** Elements of the front of the right suspension of the M51 are in view. To the far right is the dual compensating wheel, to the rear of which is the compensating-wheel link. At the center is the front shock absorber. At the lower left is the front right dual road wheel. **Above left:** The M51 had four track-return rollers per side: two less than the T43/M103. The two front road wheels and the two rear road wheels were fitted with shock absorbers. **Above right:** On the right side of the cab is the crane operator's door, right. Below the door are brackets for mounting a removable crew ladder. To the rear of it is a rack for two five-gallon liquid containers, followed by a holder for a portable fire extinguisher.

Top left: Some details of the right suspension and hull are portrayed. Two eyes are welded to the side of the hull. At the lower center is a suspension arm. Bolted to the hull between the two shock absorbers is a bumper-spring assembly; one was paired with each road wheel. **Top right:** The sprocket drum has a circumferential raised channel around the middle to engage the track guides. A vertical joint and weld in the lower hull is to the front of the sprocket. A close-up view is provided of the track connectors and the rubber track blocks. **Above left:** The left sprocket assembly is viewed from the side, showing its 11 teeth, the 11 hex screws in recesses that fasten the sprocket to the drum, and the hub located inside the drum. Three oblong openings are incorporated into the outer half of the drum. **Above right:** At the center rear of the rear deck of the M51 is a vertically oriented swivel sheave. To the immediate front of it is the travel lock for the crane. Two lugs on the bottom of the crane are secured with pins to the travel lock to keep the crane from shifting.

Left: A Recovery Vehicle, Heavy, M51, is observed from the rear, with the crane hook and the crane in the foreground and the rear spade in the raised position at the rear of the hull. Between the hydraulic cylinders of the spade are the tow pintle, the taillights, and two towing eyes. **Right:** Hanging below the rear of the crane boom is the crane hook. The hook is attached to the bottom of a multi-sheave block with coffin-shaped hexagonal sides. Here, the block is snugged-up against a multi-sheave block at the rear of the crane boom. The end of the crane cable is anchored to the rear of the boom with shackles; the other end of the cable runs through the boom to the crane-hook winch housed inside the forward part of the crane.

Left: At the top of the crane are the boom sheaves (top), below which is the upper part of the crane hook assembly, fitted with three sheaves. The multiple sheaves on the boom and on the crane hook acted as force multipliers, greatly enhancing the lifting power of the cable.

Right: The crane hook of an M51 heavy recovery vehicle is seen from the side, snuggled up against the upper part of the crane. The hook itself was very heavy duty, commensurate with the 45-ton lifting capacity of the crane.

Top left: The tail-light assemblies of the M51 are protected by arched steel guards. Below the lights are towing eyes. Between the towing eyes is a tow pintle on a heavy-duty bracket. Heavy weld beads are present of the light guards, towing eyes, and pintle bracket. **Top right:** A tow bar was stowed in cradles on each fender; this one is on the left fender. Grab fender, fitted with three hinges. In the background is the left rear of the cab. **Above:** In a view of the rear deck from the left side, at the top is the crane, and the raised steel structure to the left is the left heat deflector in the deployed position. The right heat deflector is in the background. A U-shaped tow-cable holder is on the side of the crane.

Top left: Two steel rods with eyes on the ends that are affixed to a steel beam at the rear of the cab are connected on the other end to a sheave-and-cable assembly in the crane that actuates the elevation of the crane. In the background is the cupola with a machine gun mount. **Above left:** The upper part of the crane is viewed from the crew compartment roof. On top of this part of the crane are two access pan-

Right: The top of the crane of the M51 is observed from the roof of the crew compartment. Between the steel rods is an opening for accessing the hydraulic motor inside the crane; normally, this hole would have been covered

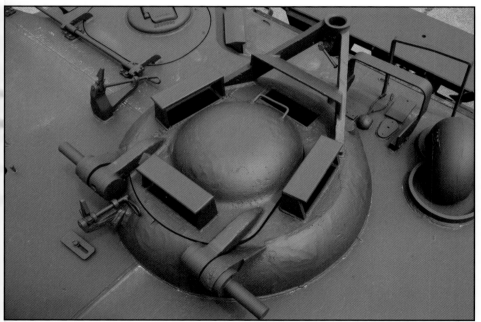

Top left: Two steel rods with eyes on the ends that are affixed to a steel beam at the rear of the cab are connected on the other end to a sheave-and-cable assembly in the crane that actuates the elevation of the crane. In the background is the cupola with a machine gun mount. **Top right:** In a view of the top of the right side of the crew compartment facing forward, in the foreground are the crane operator's vision door and his hatch, fitted with a torsion-bar lift assist. To the left of the hatch and at the front of the crew compartment roof are two rollers for a cable from the auxiliary winch. **Above left:** The commander's cupola on the M51 had a mount for a .50-caliber machine gun and provisions for four periscopes: three on the cupola rim and one at the rear of the hatch. At the left rear of the cupola was a hold-open latch for the commander's hatch. To the upper left is the driver's hatch. **Above right:** The driver's hatch on the M51 was fitted with a torsion-bar lift-assist and an M19 periscope with a hinged cover. To the front and the left side of the hatch were four M17 periscopes. At the lower center is the hold-open latch for the driver's hatch.

The M51 heavy recovery vehicle and its sibling, the T43/M103 heavy tank, signified early attempts by the U.S. armed forces to produce and deploy the types of heavily armed and armored vehicles that could survive and prevail against the increasingly formidable armored forces of the Eastern Bloc in the early years of the Cold War. Although not produced in great numbers and not tested in battle, the T43/M103 and the M51 helped pave the way for today's powerful American tanks and armored vehicles.